BRITISH GEOLOGICAL SURVEY

Geology of the Buckingham district

a brief explanation of the geological map
Sheet 219 Buckingham

M G Sumbler

Bibliographic reference

SUMBLER, M G. 2002. Geology of the Buckingham district —
a brief explanation of the geological map.
Sheet Explanation of the British Geological Survey.
1:50 000 Sheet 219 Buckingham (England and Wales).

Keyworth, Nottingham: British Geological Survey

CONTENTS

Plates

Cover photograph Now used as a museum, Buckingham Old Gaol utilises a range of building stone from almost every geological formation in the Buckingham district. (*Photograph* Paul Tod; MN39727.)

Notes

Throughout this publication, the word 'district' refers to the area covered by the 1:50 000 Series Geological Sheet 219 (Buckingham). National Grid references are given in square brackets; all lie within 100 km grid square SP unless otherwise stated; those for deep boreholes cited in the text are given in Table 1. The National Grid is also indicated on the margins of certain diagrams, based on Ordnance Survey mapping.

Acknowledgements

This Sheet Explanation was written by M G Sumbler. A J M Barron contributed to the stratigraphic account of the Inferior and Great Oolite groups, N J P Smith to that of the basement geology and M A Lewis to the hydrogeology. The contribution of the many surveyors involved in geological mapping of the Buckingham district is also acknowledged. They are (in alphabetical order): D T Aldiss, K Ambrose, C R Bristow, A J M Barron, P J Henney, P M Hopson, A Horton, A N Morigi, J Pattison, M D A Samuel, A Smith, M G Sumbler, R G Thurrell, and R J Wyatt. Dr M J Oates kindly provided unpublished information on many temporary exposures in the district.

1   Introduction

This *Sheet Explanation* summarises the geology of the district covered by 1:50 000 Series Sheet 219 Buckingham. Much of the district retains its rural character, although the towns, the largest of which is Bicester in the south-west, are expanding rapidly.

The district is divisible into four parts, each with a distinct topography acquired from the geology. The western part, underlain by Middle Jurassic strata, has much of the character of the Cotswolds, with narrow valleys cutting through open 'uplands'. The south-central part of the area is a 'clay vale' of low relief, excavated into mudstones such as the Oxford Clay. The south-eastern part is dominated by flat-topped hills capped by Portland, Purbeck and Lower Cretaceous strata. Quainton Hill [751 213], at 190 m, is the highest point in the district (Plate 1). The north and north- eastern part is essentially a plateau underlain by thick glacial drift deposits, but is dissected by valleys of the Great Ouse system.

Geological history

The oldest rocks proved in the district are Lower Palaeozoic marine strata, overlain unconformably by sedimentary rocks of Devonian to Carboniferous age. The succession was then folded during the Variscan orogeny, when the region was uplifted to become dry land. It remained a stable structural block, the London Platform, throughout the Mesozoic.

Erosion of the London Platform during Permian and early Triassic times removed much of the younger Palaeozoic succession. The oldest Mesozoic strata, continental sediments of mid to late Triassic age, rest upon a substantial unconformity. A marine transgression that began in latest Triassic times began to submerge the platform, a process that continued throughout most of the Jurassic. Shallow marine, carbonate-rich sediments were laid down, but intercalations of fluvial, deltaic or 'lagoonal' strata indicate the continued proximity of land on more easterly parts of the platform.

Uplift of the region in earliest Cretaceous times created dry land for a relatively brief period, after which fluvial sand was deposited, followed by marine sand and clay. As the sea deepened in the Late Cretaceous, a considerable thickness of Chalk was probably laid down across the region. Further uplift in Palaeogene times terminated marine sedimentation, and the region probably remained emergent throughout the Cainozoic. More than sixty million years of erosion gradually removed the Chalk and produced precursors of some of the more important features of the landscape by the early Quaternary.

The Quaternary, spanning the last two million years, was characterised by numerous cold-temperate cycles. Glacier ice probably entered the district only during the Anglian Stage, a few hundred thousand years ago, leaving thick deposits of detritus in the north. Otherwise, the district lay in the periglacial zone and was subjected to rigorous weathering. The 'post-glacial' Holocene epoch, which began about 10 000 years ago, has been a time of more gentle weathering, when stream floodplain deposition and soil formation, aided by agricultural practices, completed development of the landscape.

History of research

Prior to publication of 1:50 000 Sheet 219, the only medium-scale geological maps to cover the district were Old Series One-Inch (1:63 360) Sheets 45 and 46, published in 1863 and 1865 respectively. Sheet 45, covering the western part of the district, was described by Green (1864).

The eastern part of the district was surveyed at 1:10 560 scale (six inches to the mile) between 1895 and 1897. The north-eastern

Figure 1 Deep boreholes in the Buckingham district and environs giving depths in metres to base of unit proved. (See also Figure 2).

Name (and abbreviation on Figure 2)	BGS Reg. No.	NGR	Datum level (m OD)	Depth to base (m)							
				Middle Jurassic	Lower Jurassic	Triassic	Carboniferous	Devonian	Silurian	Cambro-Ordovician	Precambrian
Akeman Street (AS)	SP52SW/16	5207 2056	84.0	39.6	152.5	177.0	388.4	400.5	-	-	-
Bicester No.1 (B)	SP52SE/1	5878 2081	85.6	72.5	161.5	168.6	absent	341.1	508.4	-	-
Calvert West (CW)	SP62SE/2	6870 2458	67.7	45.4	115.8	absent	absent	absent	absent	197.8	-
Calvert East (CE)	SP62SE/1	6903 2457	88.4	65.2	135.2	absent	absent	absent	absent	426.1	-
Deanshanger (D)	SP73NW/2	7652 3880	69.7	63.4*	116.1	191.6	absent	absent	absent	259.1	-
GBK17	SP63SW/1	6059 3330	122.8	25.3	159.1	222.2	absent	228.6	-	-	-
GCN111	SP52SW/2	5183 2455	106.1	27.7	155.4	201.5	-	-	-	-	-
GCN116	SP52SW/4	5380 2350	81.7	32.9	139.6	172.5	187.5	-	absent	-	-
GCN160	SP62NW/2	6277 2845	90.8	42.7	139.6	150.9	absent	absent	absent	?191.1	-
Lillingstone Lovell (LL)	SP74SW/1	7197 4197	125.9	47.2	173.6	270.2	absent	absent	absent	300.2	-
Marsh Gibbon (MG)	SP62SW/1	6481 2374	75.3	39.9	121.5	absent	absent	absent	absent	174.3	-
Noke Hill (NH)	SP51SW/1	5386 1285	95.9	43.3	110.3	c.115.8	absent	247.8	-	-	-
Northbrook (N)	SP42SE/10	4994 2246	107.4	43.0	164.5	209.0	533.8	592.2	-	-	-
Steeple Aston (SA)	SP42NE/12	4687 2586	130.7	15.4	164.4	229.0	775.9	975.3	-	-	-
Tattenhoe (Ta)	SP83SW/1	8289 3437	102.4	79.2	149.3	absent	absent	absent	absent	213.2	-
Tring No.1 (Tr)	SP91SW/28	9121 1036	153.6	276.1	absent	absent	absent	absent	absent	307.2	-
Twyford No.1 (T1)	SP62NE/2	6802 2567	88.7	63.4	132.3	139.0**	absent	absent	absent	156.7	-
Twyford No.2 (T2)	SP62NE/3	6760 2650	85.0	45.1	137.8	139.8**	absent	absent	absent	153.9	-
Twyford No.3 (T3)	SP62NE/4	6859 2659	85.0	57.3	134.2	absent	absent	absent	absent	143.3	-
Twyford No.4 (T4)	SP62NE/1	6697 2561	89.0	58.8	140.2	140.8**	absent	absent	absent	150.6	-
Westcott No.2 (W)	SP71NW/4	7096 1649	78.2	124.2	163.7	absent	absent	absent	absent	173.7	-
Withycombe Farm (WF)	SP44SW/9	4319 4017	144.0	-	178.8	351.8	841.12	absent	absent	1035.1	1064.9

* In Deanshanger Borehole this figure refers to the base of the glacial drift deposits, which rest directly on Lias

** 'Twyford Beds' only, see text

part of the district was partially surveyed in the 1960s, in connection with the construction of Milton Keynes new town (Horton et al., 1974). In the early 1990s, much of the drift-covered northern part of the district was surveyed to identify possible resources of sand and gravel (see Information sources). The survey of the district was completed between 1998 and 2000.

Previous studies have concentrated on the Mesozoic rocks. The Quaternary has remained largely neglected, but research on the deep geology of the area commenced with the drilling of two deep boreholes at Calvert in the early 20th century, followed by a number of exploration boreholes drilled mainly in the 1960s (Figure 1).

Plate 1   Landslips in the Kimmeridge Clay Formation on the north side of Quainton Hill [748 223]. The hummocky topography, indicative of multiple rotational slips, is well seen in this ancient pasture land, but can be quickly obscured by modern ploughing methods (MN39725).

2 Geological description

The oldest rocks at outcrop belong to the Lower Jurassic Lias Group, but many boreholes have proved older rocks (Figure 2). The distribution of these rocks at depth is also indicated by evidence from regional aeromagnetic and gravity surveys, together with some seismic data.

Precambrian and Lower Palaeozoic

Precambrian volcanic rocks (basaltic andesite) were proved at a depth of more than 1000 m in the BGS Withycombe Farm Borehole, near Banbury, 12 km to the northwest of the district (Poole, 1978). They are overlain by Lower Cambrian sedimentary rocks (Rushton and Molyneux, 1990). Geochemically, the andesites are similar to those of Charnwood Forest in Leicestershire (Pharaoh et al., 1987); it is likely that associated Charnian rocks underlie the whole region. The **Lower Cambrian** rocks in the Withycombe Farm Borehole comprise mudstone, limestone and sandstone. Cambrian rocks may also be present at depth in the Buckingham district.

Lower Ordovician (Tremadoc) strata have been proved by a number of boreholes (Figure 2; Davies and Pringle, 1913; Bulman and Rushton, 1973). They comprise grey and greenish grey mudstone and siltstone, in all cases dipping steeply (25° to 90°). Their total stratigraphical thickness is difficult to determine because of probable folding and repetition, but the eastern borehole at Calvert penetrated 291 m of Tremadoc strata, including two thin olivine-basalt sills (Whitaker, 1921).

The Bicester No. 1 Borehole did not reach Tremadoc strata, but terminated in altered basaltic and andesitic lava and tuff inferred to be of early **Silurian** age (Pharaoh et al., 1991). Aeromagnetic data suggest that these rocks underlie the axis of the Charlton Anticline (see pp.20–21). They are succeeded in the borehole by thin, interbedded sandstone and mudstone, also probably of early Silurian age.

Upper Palaeozoic

Lower Old Red Sandstone (Early Devonian) rocks may be present in the south-western part of the district, but are overstepped by **Upper Old Red Sandstone** (ORS) (Late Devonian) farther east. The succession in the Bicester Borehole comprises horizontally bedded, purplish brown mudstone and greenish grey, fine-grained sandstone, probably mainly of fluvial origin with thin units of marine limestone. These beds may correlate with strata of probable Famennian age proved to the west and south-west in boreholes at Steeple Aston, Northbrook and Noke Hill (Poole 1977; Butler, 1981).

Carboniferous rocks are preserved at depth in the western part of the district (Figure 2; Foster et al., 1989). They occupy the eastern limb of the north–south-trending Oxfordshire Coalfield Syncline. The Carboniferous succession is nearly 1000 m thick in the centre of the syncline (Poole, 1969), but Triassic strata overstep eastwards. Only the basal 200 m or so of the succession, belonging to the **Arenaceous Coal Formation** (ACF), is likely to be present in the Buckingham district. Grey sandstone dominates the succession, with some mudstone, seatearth and coal. Dolerite sills occur in some boreholes.

Triassic and Lower Jurassic

The Worcester Basin, 25 km to the west, contains thick Permo-Triassic and Lower Jurassic successions, but subsidence was much less in the Buckingham district. Successions are consequently much thinner, with no Permian or early Triassic rocks. The Triassic and Jurassic successions in the district thin rapidly

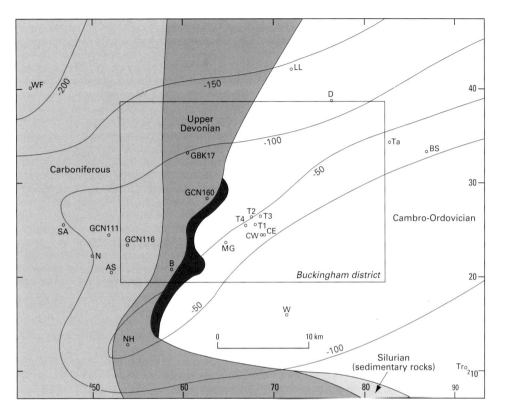

°WF
-200
-150
°LL
D
°
Upper
Devonian
-100
°Ta
°BS
° GBK17
Carboniferous
-50
GCN160
°
40 —
30 —
Cambro-Ordovician
°SA
GCN111
°
GCN116
°
T2
T4 ° °T3
° °T1
CW ∞ CE
MG
°
V
B
°
AS
°
°N
Buckingham district
W
°
20 —
NH
°
0
10 km
-50
-50
-100
Silurian
(sedimentary rocks)
Tro
°210 —
°50
°60
°70
°80 ▶
°90

Figure 2 Geological sketch map of the Palaeozoic rocks beneath the Mesozoic cover. Contours show the level of the surface in metres relative to Ordnance datum (OD). Location of deep boreholes also shown (see Figure 1 for details).

V Volcanic rocks of probable Silurian age.

south-eastwards, as individual units are condensed and older units are overlapped by younger ones.

The **Sherwood Sandstone Group** occurs at depth in the north-west of the district, but is represented only by the **Bromsgrove Sandstone Formation** (BmS) of Mid Triassic age (Figure 3; Horton et al., 1987). The formation is probably up to about 35 m thick in the north-west, but thins south-eastwards and is eventually overlapped by younger Triassic and Jurassic beds. It is absent from

the boreholes at Bicester, Twyford and Deanshanger. The formation comprises brown, fine to medium-grained, more or less argillaceous sandstone and siltstone, with some mudstone beds, and is dominantly of fluvial origin.

Up to 73 m of the Mid to Late Triassic Mercia Mudstone Group (MMG) have been proved in the north-west of the district, but it wedges out a few kilometres south-east of the Bromsgrove Sandstone Formation, overlapping the latter to rest on the Palaeozoic

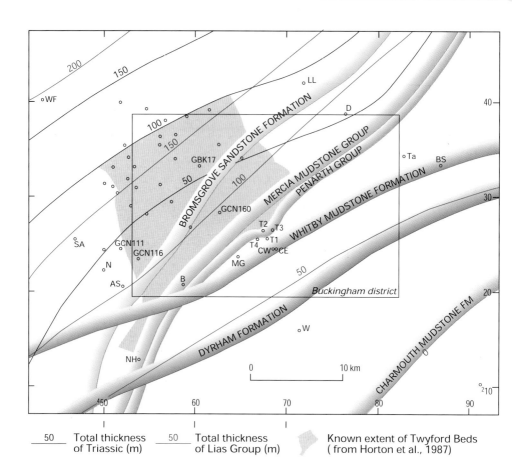

Figure 3 Sketch map showing approximate limits of Triassic and Lower Jurassic formations beneath younger deposits, and approximate total thicknesses of the Triassic and the Lias Group.

basement (Figure 3; Sections 1, 2 on Sheet 219). The group comprises mainly reddish brown mudstone and siltstone that accumulated in playa lakes in a desert environment. The topmost beds are commonly greenish grey, and for this reason have been ascribed to the Blue Anchor Formation. However, it is more probable that they are older beds, originally red, that have been altered by chemical reduction processes associated with the marine conditions of the

overlying Penarth Group (Horton et al., 1987).

The **Penarth Group** (PnG) is a thin succession of marine strata forming the youngest (Rhaetian) part of the Triassic and resting on an eroded surface of the Mercia Mudstone Group. Where complete, it comprises dark grey marine mudstone of the **Westbury Formation**, overlain by the paler mudstone of the **Cotham Member** and the limestone-dominant **Langport Member** ('White Lias'), both of the **Lilstock Formation**. This succession is probably present in the western part of the district, where the group is up to 7 m thick, but these lithologies are absent farther east, where greenish grey siltstone and calcareous, locally conglomeratic sandstones form the '**Twyford Beds**'. The latter were originally proved by Twyford boreholes 1, 2 and 4, where they rest directly on Palaeozoic strata (Donovan et al., 1979). According to Horton et al. (1987), they represent a near-shore facies of the Penarth Group. This may be correct, although there is no biostratigraphical evidence to confirm a Rhaetian age. Their lateral relationships (e.g. the limits shown on Figure 3) are also uncertain, so it is equally possible that strata assigned to the Twyford Beds may represent the basal part of the Lias Group.

The Lower Jurassic **Lias Group** (Li) is over 500 m thick in parts of the Worcester Basin, but is much thinner in the Buckingham district (Horton et al., 1987, fig. 15), where only the higher part (Sinemurian to Toarcian) is present because of onlap onto the London Platform (Donovan et al., 1979). The Lias thins dramatically south-eastwards, partly because of onlap but mainly because of erosion prior to deposition of the Inferior Oolite and Great Oolite groups (Figure 3). Only the highest part of the Lias (Whitby Mudstone Formation) reaches outcrop within the district, although the older Charmouth Mudstone Formation is present at rockhead in a deep, drift-filled channel proved by the Deanshanger Borehole [765 380 area].

The **Charmouth Mudstone Formation** (ChM) corresponds approximately to the Lower Lias of previous accounts (e.g. Horton et al., 1987; 1995), and to the Brant Mudstone of Shephard-Thorn et al. (1994). It is dominated by grey mudstone with sporadic thin beds and nodules of argillaceous limestone. Ammonite biozones have been established in a few cored boreholes, and downhole gamma ray and electrical logs have been used to correlate many uncored boreholes in and around the district (see Horton et al., 1987). The beds at the base of the succession commonly have a 'spiky' gamma-ray signature, which suggests interbedded limestone and mudstone, and represents a diachronous shallow-water facies that becomes younger to the south-east due to onlap.

The Charmouth Mudstone is up to about 130 m thick in the north-west, but thins to an estimated 25 m or so in the south-east corner where it rests directly on the Palaeozoic. Part of this thinning is due to loss of the basal beds by overlap, the base of the formation being of Turneri Zone (Sinemurian) age in the Steeple Aston Borehole, west of the district, and Jamesoni Zone (early Pliensbachian) age in the Tattenhoe Borehole, east of the district.

The **Dyrham Formation** (DyS), corresponds approximately to the Middle Lias Silts and Clays of previous accounts. It consists of grey, silty, finely micaceous mudstone, mainly of late Pliensbachian (Margaritatus Zone) age. The lower boundary with the Charmouth Mudstone is gradational, but can be recognized on borehole gamma-ray logs in which the more silty mudstones of the Dyrham Formation have generally lower counts than the underlying strata. Where complete, the formation ranges from about 15 m thick in the north-west to between 6 and 8 m in the Calvert and Twyford boreholes. Farther southeast, it is cut out by the Great Oolite Group.

As with the underlying units, there is a pronounced south-eastward thinning of the **Marlstone Rock Formation** (MRB), from up to 7.5 m thick near Banbury, to less than 4 m in boreholes near the north-west corner of the district, and more generally about 1.8 m, for example at Calvert and Twyford. Its most south-easterly proving was in the Calvert 1/76 Borehole [6888 2462], in which only the uppermost 0.33 m was proved, comprising brown and greenish grey, ferruginous, sandy, shell-fragmental and ooidal limestone. The formation is probably entirely of late Pliensbachian (Spinatum Zone) age in the district. It is probably cut out by the Great Oolite Group towards the south-east.

The **Whitby Mudstone Formation** (WhM), the Upper Lias of previous accounts, is dominated by dark grey silty and finely micaceous mudstone. It is up to 38 m thick in the north-west, but thins south-eastwards to between 3 and 4 m at Calvert before being cut out by the Great Oolite Group (Figure 3). The basal 'Fish Beds', generally only a few centimetres thick, are dark fissile mudstone. They are succeeded by up to 1 m or so of beds containing nodular, often ammonite-rich 'Cephalopod Limestones'. Overlying beds are grey mudstone with sporadic limestone nodules. The strata belong to the Toarcian Falciferum and Bifrons zones.

Middle and Upper Jurassic

The **Inferior Oolite Group** is over 100 m thick in the north Cotswolds, but is cut out eastwards by the Great Oolite Group. It is very thin in the Buckingham district, being represented only by the **Northampton Sand Formation** (NS) of Aalenian age. In the Ardley–Fritwell railway cutting [514 286] (Figure 4), the Northampton Sand Formation comprises hard, massive to rubbly, bluish grey to brown calcareous sandstone, weathering to reddish brown and decalcifying locally to brown sand (see Arkell et al., 1933, in which the formation is misidentified as 'Hook Norton Beds'). In boreholes, the formation is commonly described as 'hard sandstone', in contrast to the weakly cemented sands above. It is generally 1 to 3 m thick, although locally up to 6 m, as in the Ardley–Fritwell cutting. It does not extend very far to the east, probably being cut out by the Horsehay Sand (Figure 4) along a line between Brackley and Bicester.

The **Great Oolite Group** (GtO) is up to about 45 m thick in the west of the district, thinning eastwards to perhaps 20 m. Most of the thinning takes place in the lower part of the group. Overall, the sedimentary facies are transitional between the fully marine succession of the Cotswolds and the paralic/non-marine environments of the East Midlands. Facies changes across the district are particularly evident in the lower part of the group, in the Horsehay Sand, Sharp's Hill, Taynton Limestone and Rutland formations, all of which pass north-eastwards into an extended Rutland Formation (formerly known as the Upper Estuarine Series; Figure 5). The whole of the succession is represented in the Ardley–Fritwell railway cutting (Figure 4),

between Bucknell [558 250] and Fritwell Tunnel [514 291].

The sandy strata overlying the Northampton Sand Formation are now termed the **Horsehay Sand Formation** (HS) (Cox and Sumbler, 2002). They have been attributed in the past to both the Inferior Oolite Group (i.e. the Grantham Formation of the East Midlands) and the Great Oolite Group. Bradshaw (1978) argued that they belonged in the latter, and represented a south-westward transition from the fluviolacustrine Stamford Member, at the base of the Rutland Formation in Lincolnshire, to the fully marine Chipping Norton Limestone of Oxfordshire. This view is supported by palynological evidence (Fenton et al., 1994, 1995). The beds were deposited in a marginal marine, perhaps deltaic environment (Sellwood and McKerrow, 1974).

The Horsehay Sand is represented in the Ardley–Fritwell cutting by several metres of varicoloured, unconsolidated sands that channel deeply into the Northampton Sand Formation. Elsewhere, the formation also includes beds of siltstone and mudstone (Figure 4). Rootlets and other plant material are locally common, but other fossils are rare. The top of the formation is typically an erosional non-sequence and may be leached, with many rootlets indicating establishment of a vegetated swamp environment. The formation can be recognized in boreholes, in which it is typically 5 m or so thick with a probable maximum of about 7 m, although the position of the upper boundary may be uncertain in some boreholes because of the presence of mudstone beds (compare with the Sharp's Hill Formation). The mudstones seem to become more dominant in the east (e.g. the Tattenhoe borehole; Figure 4), in parallel with the lateral passage into the non-marine

Figure 4   Boreholes and exposures in the Great and Inferior Oolite groups, showing the eastward thinning of the succession across the district. Inset shows location of sections.

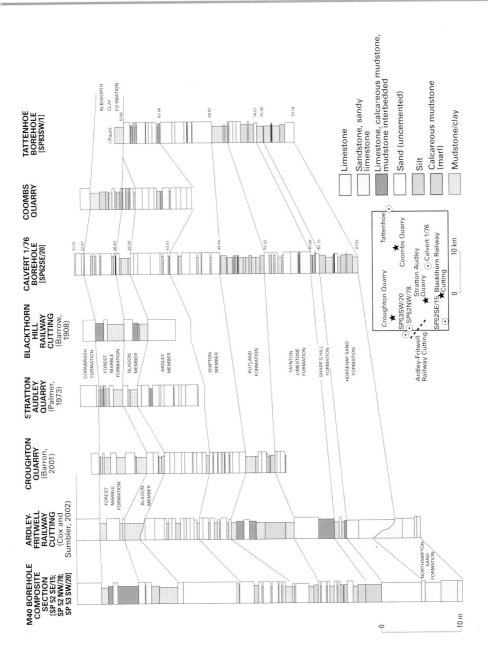

Figure 5 Lithostratigraphical nomenclature and correlation of the Great and Inferior Oolite groups in the Buckingham District and adjoining areas. Not to scale.

Chipping Norton district (Sheet 218)	Buckingham district (Sheet 219)		Towcester–Bedford–Northampton districts (Sheets 202, 203, 185)	
Cornbrash Formation	Cornbrash Formation		Cornbrash Formation	
Forest Marble Formation	Forest Marble Formation		Blisworth Clay Formation	
White Limestone Formation	White Limestone Formation	Bladon Member		
		Ardley Member	Ardley Member	Blisworth Limestone Formation
		Shipton Member	Roade Member (Sharpi Beds)	
Hampen Formation	Rutland Formation		(un-named unit)	Rutland Formation
Taynton Limestone Formation	Taynton Limestone Formation		Wellingborough Limestone Member	
Sharp's Hill Formation	Sharp's Hill Formation		(un-named unit)	
Chipping Norton Limestone Formation	Horsehay Sand Formation		Stamford Member	
Leckhampton Member of Birdlip Limestone Formation ('Scissum Beds')	Northampton Sand Formation		Northampton Sand Formation	

Stamford Member. There is little evidence from the south-eastern part of the district, but it is likely that the unit is present there, although perhaps only 1 to 2 m thick.

The **Sharp's Hill Formation** (ShH) is a thin unit of mudstone and silty mudstone with subordinate limestone. In contrast to the Horsehay Sand, it is fully marine and generally contains abundant fossils, particularly oysters and gastropods. In the Ardley–Fritwell cutting, the basal part of the Sharp's Hill Formation includes a highly carbonaceous bed, named the 'Peat Bed' by Arkell et al. (1933). Similar lithologies, commonly described as 'peat' or 'coal', occur at this level in boreholes and water wells elsewhere in the district; they probably represent the remains of swamp vegetation that grew on top of the Horsehay Sand delta before inundation by the sea. The Sharp's Hill Formation is up to

4 m thick, but is more generally 1 to 2 m, with some indication of thinning in the eastern part of the district.

The **Taynton Limestone Formation** (Ty) consists of cross-bedded, shell fragmental, ooidal limestones (grainstones) with subordinate finer grained limestones (packstones), generally with some interbeds of calcareous mudstone and mudstone. In wells and boreholes, the Taynton Limestone is recorded as 'hard grey rock' or 'grey stone', and commonly noted as water-bearing. Thicknesses range from 0 to 7 m, with much local variation because of the development of argillaceous lithologies that are classified with the Rutland or Sharp's Hill formations. Overall, the formation thins towards the east and south-east.

The formation is some 5 m thick in the Ardley–Fritwell cutting (Arkell et al., 1933, beds 12 and 13). The upper part is cross-bedded oolite, but the lowest 2 m or so comprise interbedded argillaceous limestone and calcareous mudstone, similar to the laterally equivalent Wellingborough Limestone Member of the Rutland Formation farther north-east (Figure 5).

The **Rutland Formation** (Rld) is dominated by mudstone, calcareous mudstone, siltstone and fine-grained sandstone, with very minor limestone. The formation has been termed Hampen Marly Beds in previous accounts of the region (e.g. Arkell et al., 1933; Palmer, 1979, Horton et al., 1987), but the latter, now the Hampen Formation, is substantially different, being dominated by fully marine limestones in its type area (Sumbler and Barron, 1996). The Rutland Formation passes into the Hampen Formation a few kilometres west of the Buckingham district (see Horton et al., 1995, fig. 8). The Rutland Formation of the Buckingham district correlates with only the upper part of the type succession in the East Midlands (Figure 5).

The Rutland Formation ranges from about 2 to 12 m in thickness, with the thinnest successions in the east and south-east. Part of the variation results from the arbitrary boundary with the White Limestone, the lithologies of which have a gradational and interdigitating relationship with those of the Rutland Formation. Arkell et al. (1933) recorded about 8.4 m of Rutland Formation in the Ardley–Fritwell cutting. The lowest 5.6 m comprise green and dark grey sandy clay, with some beds of sandy, muddy limestone. The basal bed (Bed 14) has yielded marine bivalves and brachiopods, and the holotype of the ammonite *Procerites imitator* (S S Buckman) was collected from just above (Bed 15). Marine faunas occur sporadically in higher parts of the Ardley–Fritwell section, but much of the succession comprises blocky mudstone and siltstone that is unfossiliferous except for carbonaceous plant material. According to Palmer (1979), the formation was deposited in a near-shore region of shallow brackish water lagoons, with rootlet beds indicating the occasional development of saltmarsh conditions.

The **White Limestone Formation** (WhL) is moderately well exposed, notably in the Ardley–Fritwell cutting [520 285 to 548 261], the adjacent Ardley Quarry (Plate 3), the Blackthorn Hill railway cutting [617 211], Stratton Audley Quarry [601 252] and Coombs Quarry [733 327] (Plate 2). The formation is dominated by white, cream and buff limestone (wackestone), with varying proportions of peloids (mainly faecal pellets) and shell debris set in a micrite matrix. Beds of grey marly limestone and mudstone occur at some levels, but higher energy packstones and grainstones are rare. The formation is divided into the Shipton, Ardley and Bladon members (Sumbler, 1984, emending Palmer, 1979), separated by marker beds. The members can be readily identified in exposures and generally also from borehole records. However, the Shipton and Ardley members cannot usually be mapped separately and are shown as White Limestone Formation undivided on the map. In contrast, the Bladon Member, where present, has been separated because of its distinctive lithologies (but see below).

The **Shipton Member** varies between about 3 and 7 m in thickness. It comprises peloidal wackestone and packstone with beds of argillaceous limestone and variably calcareous mudstone. The mudstone varies in abundance from place to place, and the boundary with the underlying Rutland Formation is one of lateral passage and interdigitation, accounting for the bulk of the thickness variation. The top of the member is marked by the Excavata Bed, the upper part of which may be developed as a strongly cemented hardground (indicative of submarine lithification) capped by an erosion surface with borings and encrusting oysters. Dinosaur footprints have been found at this level in Ardley Quarry (Plate 3), implying a shallow, perhaps intertidal environment. The **Ardley Member** ranges from about 5 to 10 m in thickness, in general thinning eastwards. It is made up almost entirely of peloidal wackestone, although a bed of sandy mudstone or siltstone occurs at the base and forms a useful marker in boreholes (including gamma-ray logs). It passes up into a sandy limestone known as the Roach Bed. The topmost bed of the member is the Bladonensis Bed, the top of which is typically a wackestone that may

Plate 2   Great Oolite Group, Coombs Quarry [733 327]. The geological trial in this old quarry shows the Ardley and Bladon members of the White Limestone Formation (forming the steep lower face), the Forest Marble (the prominent limestone bed at the top of the lower face, and the thick overlying clays) and the Cornbrash Formation (forming the subsoil at the very top of the section). See also Figure 4 (MN39726).

show lamination, shrinkage cracks and clay-filled fissures or rootlets. The **Bladon Member** (Bl) varies from 0 to about 5 m or more in thickness; 1 to 3 m is typical (Plate 4). The lower part, the Fimbriata–Waltoni Bed, comprises calcareous mudstone and greenish grey to black mudstone, commonly containing much lignite. It is believed to represent a nearshore saltmarsh environment. The upper part of the Bladon Member, the Upper Epithyris Bed, comprises marine limestones that are generally more or less marly, white-weathering wackestones with few clasts, particularly in the west (the 'Cream Cheese Bed' of Odling, 1913).

The **Forest Marble Formation** (FMb) is generally 3 to 4 m thick, but varies from about 2 to 7 m. The thickness variations result mainly from its erosive base, which locally cuts out much or all of the Bladon Member (Sumbler, 1984), but may be compounded by inclusion of the latter; the mudstones and calcareous mudstone of the Bladon Member are much like those of the Forest Marble Formation and may have been mapped with the Forest Marble in places. The Forest Marble of the district is dominated by grey mudstone and greenish buff marl (Plate 2). Units of limestone occur sporadically and randomly, and may make up

Plate 3   Dinosaur Trackway, Ardley Quarry [543 255]. The top surface of the Shipton Member at Ardley Quarry shows a variety of fossil footprints testifying to the extreme shallowness of the sea at the time of formation. This spectacular 'dinosaur trackway' was probably made by a sauropd such as *Cetiosaurus*, bones of which have been found in the White Limestone at several localities in the region. The footprints are approximately 60 cm across (MN39729).

the greater part of the succession at some localities (Figure 4). The limestones are mostly channel-fills. They are predominantly flaggy, shell-fragmental, ooidal, commonly sandy or argillaceous limestones, and may contain fossil bivalves, notably small oysters (*Praeexogyra hebridica*). Lower energy, finer grained lithologies also occur, and many of the rocks are recrystallised, making internal structure obscure. Limestones tend to be thinner and rarer in the north-east. To the north and east of the district, the Forest

Marble passes into the upper part of the Blisworth Clay Formation (Figure 5), representing a nearshore and marsh environment like that of the Rutland Formation.

The **Cornbrash Formation** (Cb) consists of medium to fine-grained, shell-fragmental wackestone and packstone with rare peloids, with subordinate thin beds of calcareous mudstone and clay. The rock is intensely bioturbated and consequently poorly bedded. It is bluish grey where fresh, but weathers to an olive or yellowish brown and produces a platy brash in arable fields. It generally rests erosively on mudstone of the Forest Marble. The formation ranges from about 1 to 4 m in thickness, and is generally very close to 3 m. Apparently greater thicknesses (over 4 m) in some boreholes may include limestone of the Forest Marble Formation.

Douglas and Arkell (1932) suggested that the section at Stratton Audley Quarry [601 252] is typical for the Bicester area. The section there showed 2.36 m of hard grey limestone with a 0.53 m bed of calcareous mudstone in the upper part. The Cornbrash is moderately well exposed in Blackthorn Hill railway cutting (Figure 8) and Coombs Quarry (Plate 2).

Brachiopods and rare ammonites have been used to subdivide the Cornbrash into lower and upper parts, belonging to the Bathonian and Callovian stages respectively. Only Lower Cornbrash was recorded in the region by Douglas and Arkell (1932), but Upper Cornbrash occurs locally, for example near Foscote [718 352]. There it comprises 0.32 m of calcareous mudstone and limestone, lithologically very similar to the underlying Lower Cornbrash (Cox et al., 1991).

The **Ancholme Group**, about 150 m thick, consists predominantly of mudstone, and comprises, in ascending order, the Kellaways, Oxford Clay, West Walton, Ampthill Clay and Kimmeridge Clay formations. Cementstone (fine-grained argillaceous limestone) bands and nodules occur at many levels, but are rarely seen in the soil. They may form minor topographical features, and some are important markers in boreholes and exposures. Bivalves and ammonites dominate the fauna.

Plate 4   The uppermost part of the White Limestone Formation at Stratton Audley Quarry [601 252]. The steep lower face, exposing the upper 2.5 m of the Ardley Member, is overlain by the variegated, lignite-rich clay of the Fimbriata Waltoni Bed and pale calcareous mudstone and limestone of the Upper Epithyris Bed, which together constitute the Bladon Member (see Figure 4), nearly 2 m thick in this part of the quarry. The overlying Forest Marble (3 m) and Cornbrash (about 2 m) are obscured by tipped debris (GS1188).

These fossils are very susceptible to weathering, and generally only the calcitic belemnites and the oyster-like *Gryphaea* are to be found in the soil. The latter show substantial evolutionary changes through the Ancholme Group, and their morphology can indicate the approximate horizon.

The **Kellaways Formation** (Kys), 5 to 6 m thick, is divisible into two members, mapped separately in the south-west. The **Kellaways Clay Member** (KlC) consists of dark bluish or purplish grey, smooth-textured, slightly silty, fissile, somewhat pyritic mudstone. The junction with the overlying Kellaways Sand is gradational, and so the thickness of the unit as recorded in boreholes is variable, ranging from 1 to 4 m. Mapping suggests that a thickness of about 3 m is typical. The **Kellaways Sand Member** (KlS) forms a slight scarp on the south-eastern flank of the Langford Brook valley in the south-west, and accounts for most of the area of the Kellaways Formation outcrop elsewhere in the district. It is composed of very fine-grained, generally uncemented quartz sand, silt and mudstone. The thickness of the member ranges from 2 to 5 m. Thickening occurs mainly at the expense of the Kellaways Clay.

The **Oxford Clay Formation** (OxC) is divided into the Peterborough, Stewartby

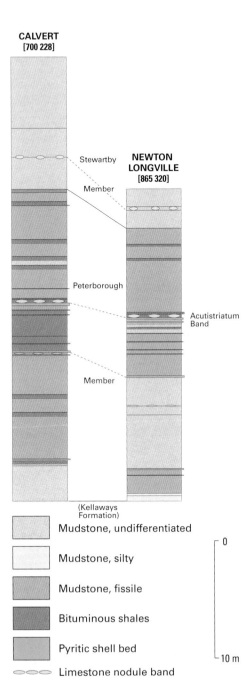

CALVERT
[700 228]

Stewartby
Member

NEWTON
LONGVILLE
[865 320]

Peterborough

Acutistriatum
Band

Member

(Kellaways
Formation)

Mudstone, undifferentiated

Mudstone, silty

Mudstone, fissile

Bituminous shales

Pyritic shell bed

Limestone nodule band

0

10 m

and Weymouth members (Cox et al., 1992). Its thickness ranges from about 67 m in the west to about 62 m in the east. The **Peterborough Member** (Pet), about 24 to 26 m thick, consists of interbedded greenish grey, slightly blocky mudstone and brownish grey, fissile, bituminous, shelly mudstone. The latter contributes to a soil that is typically darker brown and more crumbly than that developed on other parts of the formation. The only fossils to be found at surface are small *Gryphaea dilobotes*, large cylindroteuthid belemnites and, in the upper part, the 'knobbly' serpulid worm tube *Genicularia vertebralis*. Beds of grey septarian cementstone nodules occur at several levels (Horton et al., 1995). Although none have been mapped at outcrop, their levels can be inferred from the gamma-ray logs of boreholes. Only the Acutistriatum Band, which includes huge nodules commonly 1 m or more in diameter, seems to be laterally persistent, being present in the former brickpits at Calvert [700 228] (Plate 5) and, just east of the district, at Newton Longville [865 320] (Figure 6). Much of the upper part of the member (including the Acutistriatum Band) is still visible at these sites.

The **Stewartby Member** (Stw), about 21 to 23 m thick, consists predominantly of pale to medium grey, smooth or slightly silty, calcareous, blocky mudstone. It lacks the bituminous mudstone of the Peterborough Member. The member is generally poorly fossiliferous, although beds packed with immature bivalves (*Bositra*) occur, and small ammonites are common at some levels, mostly preserved as uncrushed pyrite casts. The lowest few metres of the member are exposed at Calvert and Newton Longville (Figure 6). The mudstone weathers to produce a dull, mid-greyish brown soil, much heavier and stickier than that of the Peterborough Member. The Stewartby

Figure 6   Oxford Clay successions at Calvert and Newton Longville brick pits, showing the eastward thinning of the succession (based on Callomon, 1968).

Plate 5   Oxford Clay Formation, Calvert Brickpit [700 228]. A remaining face in the once-extensive pit shows about 4 m of fissile mudstone in the uppermost part of the Peterborough Member of the Oxford Clay Formation. The 'cementstone' nodules of the Acutistriatum Band (up to 0.4 m thick) occur at the top of the face. The argillaceous limestone is detrimental to the brickmaking process, and so large quantities of discarded nodules have been dumped on the bench above (GS1189).

Member forms subdued, rather flat-topped scarps or hills capped by more resistant strata in the upper part of the member, including beds of nodular cementstone or calcareous siltstone. *Gryphaea lituola*, infilled with soft, cream limestone, is fairly common in the soil at this level. The topmost bed of the member (and of the Callovian Stage and Middle Jurassic Series) is the Lamberti Limestone, a pale grey to cream, marly limestone about 0.3 m thick. It contains many bivalve and gastropod fossils as well as a variety of ammonites, including the zonal ammonite *Quenstedtoceras lamberti*. This bed was formerly exposed at Woodham Brickpit [709 184], now infilled, just south of the district (Horton et al., 1995), and may be brought to the surface by deep ploughing or ditching, as at Graven Hill [591 205], Windmill Hill [662 246], Doddershall House [713 200] and near Winslow [760 273].

The **Weymouth Member** (Wey), some 15 to 20 m thick, consists of pale grey, smooth to slightly silty mudstone, with sporadic harder bands. It forms a grey and fawn mottled clay soil, in which *Gryphaea dilatata* and small *Hibolites* belemnites may be found. The *Gryphaea dilatata*, shows a progressive change from small, fairly tightly incurved forms transitional to *G. lituola* in the lower part of the member, to larger, broader, more open forms in the upper part. The lower half of the member was formerly exposed at Woodham Brickpit (Callomon, 1968; Horton et al., 1995).

The **West Walton Formation** (WW) comprises about 10 to 15 m of dark grey, silty mudstone with interbedded pale grey, smoother, more calcareous mudstone. The dark mudstone typically contains finely divided white shell debris and black specks of plant material. It produces a stiff grey clay soil that has dark brown and ochreous weathering colours, unlike that produced by the underlying Weymouth Member or the succeeding Ampthill Clay Formation. Race pellets (tufaceous calcium carbonate) and strings of tiny selenite crystals are particularly abundant in the subsoil. Large, thick-shelled *Gryphaea dilatata*, commonly bored and encrusted by serpulids, are abundant in some beds, and *Gryphaea*-rich beds in the upper part of the formation tend to form a shelf that caps the scarp-like outcrop. At several localities between Swanbourne [801 273] and Hoggeston [807 251], pale grey silty bands and nodules of silty cementstone, generally containing abundant specimens of the small oyster *Nanogyra*, form lenticular 'reefs' at or very close to the top of the formation, marked 'ls' on Sheet 219.

The **Ampthill Clay Formation** (AmC) is about 19 m thick in the Folly Farm Borehole [7958 1904], just outside the district (Sumbler, 1989). Thickness estimates from outcrop range from about 10 to 20 m, but much of the apparent variation may be due to superficial structural effects (pp.20–22). The formation is dominated by pale grey and mid bluish grey mudstone, which weathers to form a greyish brown clay soil. Small cementstone nodules occur at some levels; they are generally less than 15 m across, pale grey, white or yellow-weathering, smooth-textured, commonly septarian. Large *G. dilatata*, some bored and serpulid-encrusted, are common in the lower part of the formation, and occur amongst ditch debris at many localities. They are replaced by the flatter, thinner-shelled *Deltoideum delta* in the upper part of the formation. At the top of the formation, a unit of pale grey clay with cementstone, abundant *D. delta* and some large *Nanogyra nana* is found at several localities, notably near Hoggeston [818 257; 807 249].

The **Kimmeridge Clay Formation** (KC) is dominated by grey mudstone that weathers to a brown and ochreous clay, generally darker than the soils of the Ampthill Clay. Small chips and nodules of black phosphatic material are found at the base of the formation and at a few other levels, and fragments of cementstone are also found in places; they are brown-weathered and silty in texture, and generally contain white aragonitic shell fragments. At least three cementstone nodule bands are present, estimated to be about 5, 10 and 15 m above the base of the formation. The thickness of the formation is estimated to vary at outcrop from about 20 to 40 m, but most of this variation is probably due to cambering of the succeeding Portland Formation. The true thickness is thought to be within the range 35 to 40 m. This is thinner than in the district to the south (up to 56 m; Horton et al., 1995), suggesting that overall the sequence is condensed and that, in places, the highest

beds may have been removed beneath the erosive base of the Portland Group.

The **Portland Group** is very thin compared with its development in Dorset, and is represented solely by the Portland Formation. The base of the group is a significant non-sequence on top of the Kimmeridge Clay, and is commonly marked by a line of springs and seepage. The **Portland Formation** (Pl), about 12 m thick, comprises marine limestone and calcareous sand, containing a · rich fauna of large bivalves and sporadic giant ammonites (Wimbledon, 1980). The formation is divided into a Portland Sand Member overlain by a thinner Portland Stone Member (Figure 7). There are now few exposures, but published accounts (Bristow, 1963, 1968; Bristow and Kirkaldy, 1962; Davies, 1897; Fitton, 1836; Mitchell, 1834; Wimbledon, 1974; Woodward, 1895) enable comparison with the more extensive outcrops to the south (Horton et al., 1995).

The basal unit of the **Portland Sand Member** (PlS), the Upper Lydite Bed, is a thin and lenticular development of argillaceous sand, calcareous mudstone and limestone with rare creamy micritic limestone. It contains abundant coarse glauconite grains and tiny pebbles of black chert ('lydites'), white or yellowish quartz and rare black phosphate. These pebbles are conspicuous in the soil, associated with lumps of limestone that contain bivalves and fragments of the zonal ammonite *Glaucolithites*. Above, the succession is dominated by argillaceous, generally fine- to medium-grained sand. This is markedly glauconitic in the lower part and contains nodular grey to brown, coarse-grained sandy limestone (Glauconitic Beds). The upper 1 to 3 m or so are non-glauconitic, fine-grained and rarely cemented (Crendon Sand). Between the Glauconitic Beds and Crendon Sand, a variable development of sandy and shell fragmental limestone, commonly containing large bivalve fossils such as *Protocardia* and *Pleuromya*, corresponds with the Aylesbury (or Rubbly) Limestone of areas to the south (Horton et al., 1995). Locally, notably just to the north of Whitchurch [802 218 area], the Aylesbury Limestone (and, by implication, the Crendon

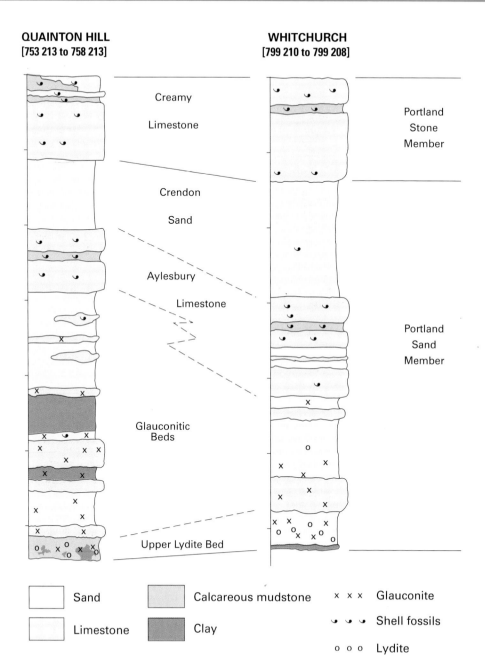

Figure 7   Portland Formation, based on trench sections recorded by Bristow (1963).

Sand) may be mapped with the Portland Stone.

The **Portland Stone Member** (PlSt) forms a minor scarp rising above the Portland Sand outcrop. It is rarely exposed, but produces abundant brash in arable fields. Typical rock types are white to yellowish, strongly burrowed, variably shell-fragmental and peloidal wackestone and packstone, commonly containing large bivalves (*Protocardia*, trigoniids), serpulids and gastropods (*Aptyxiella*, *Pleurotomaria*), and giant ammonites (such as *Titanites*). 'Roach' lithologies, in which shells are dissolved away to leave empty moulds, occur sporadically. Near the top of the succession, peloidal wackestones and calcarenites with large, flat oysters ('*Ostrea*' *expansa*) are common. Sections may include oyster-rich marls.

The **Purbeck Group** is also represented by a single formation, comprising a thin succession of limestone and calcareous mudstone. These were deposited in nearshore, shallow waters under conditions of low or fluctuating salinity. Consequently, the fauna is generally sparse, although ostracods are common at some levels (Barker, 1966; Anderson, 1985). Combined with other evidence, they suggest that the **Purbeck Formation** (Pb) of the district correlates with part of the latest Jurassic Lulworth Formation of Dorset, or even part of the underlying Portland Group there, suggesting diachronous facies change (Wimbledon, 1980; Horton et al., 1995).

The Purbeck outcrop is limited to small areas around the margins of the Whitchurch Sand. The discontinuous outcrop is mainly a result of channelling at the base of the Whitchurch Sand and consequent removal of the Purbeck, but some Purbeck may have gone undetected where obscured by sandy wash from the overlying strata. Probably up to about 2 m of Purbeck strata are present locally on Quainton Hill, and about 3 m at Oving (Fitton, 1836; Woodward, 1895; Davies, 1897). Barker (1966) recorded a composite section in trenches [around 805 225] north of Whitchurch, which proved at least

3.6 m (beds 7 to 16) and possibly over 6 m (beds 7 to 26).

The base of the Purbeck Formation is marked by the distinctive 'Pendle', which conformably succeeds the Portland. It is a flaggy limestone 0.3 to 0.5 m thick, consisting of thin (5 to 10 mm) alternating layers of grey, ostracod-rich limestone, and cream, pure lime mudstone (micrite). The overlying beds are dominated by calcareous mudstone and clay, with white and brown mottled lime mudstones. The latter may be recrystallised, hard and splintery, suggesting subaerial emergence as proved in the area to the south (Horton et al., 1995). The limestone may contain strings of ooids, as well as ostracods, small gastropods and thin-shelled bivalves.

Lower Cretaceous

Ferruginous sands above the Portland and Purbeck beds were classified as Lower Greensand on early geological maps, but most of these deposits are now assigned to the **Whitchurch Sand Formation** (WhS), an older deposit that is mostly of non-marine, mainly fluvial origin. Although a latest Jurassic age has been claimed (Casey and Bristow, 1964; Morter, 1984), several lines of evidence suggest that it is earliest Cretaceous (?Valanginian), representing part of the **Wealden Group** of the Wessex Basin (Horton et al., 1995).

The Whitchurch Sand, estimated to reach a thickness of 6 m, rests sharply on the underlying Purbeck Formation, cutting through it in places to rest on the Portland Formation. The outcrop is characterised by reddish brown, sandy loam soils, locally with a brash of hard, dark brown to purple-black, limonite-cemented sandstone (including hollow 'box-stones'). Sections show white, yellow, orange and brown, fine- to medium-grained sand, with seams of limonite-cemented sandstone. Mudstone beds are common; they are typically pale grey to almost white, with ochreous yellow mottles, and commonly contain limonitic concretions (probably altered siderite). The mudstones are known to be rich in kaolinite (Allen and Parker, *in* Horton et al., 1995).

Some records describe a black (?carbonaceous) clay at the base (Barker, 1966, Bed 27; Bristow, 1963, 1968).

Both mudstone and sandstone may contain plant material, and Fitton (1836) recorded the fresh-water molluscs 'Cyclas' (i.e. Neomiodon) and Paludina (i.e. Viviparus). Casey and Bristow (1964) recorded a more diverse marine fauna at Quainton Hill [748 221] and Whitchurch [796 210], with Protocardia and Laevitrigonia as well as other bivalves, gastropods and serpulids, suggesting either an estuarine environment or a marine incursion that may indicate a link with the probably contemporaneous Spilsby Sandstone of Lincolnshire. A ferruginous ooidal bed found north of Whitchurch [806 224] may also indicate a marine origin, although similar lithologies are known from apparently non-marine beds near Oxford (Horton et al., 1995).

Two small outcrops of the **Lower Greensand Group** (LGS) occur at Hogshaw Hill [746 218; 748 221]. They appear to fill a channel or scour cut into the Whitchurch Sand. The strata comprise reddish brown, very coarse-grained, locally conglomeratic, limonite-cemented sandstone. Casey and Bristow (1964) described excavations in the southernmost outcrop, showing 1.2 m of sandstone resting on a bored surface of cemented Whitchurch Sand. The Lower Greensand at this locality yielded abundant moulds of marine fossils, including sponges, brachiopods, bivalves (including the rudist Toucasia lonsdalei) and the ammonite Parahoplites. The beds are believed to equate with the Seend Ironsand of Wiltshire, and with the lower part of the Woburn Sands in Bedfordshire. Scattered coarse-grained or pebbly sandstone found elsewhere on the Whitchurch Sand outcrop [e.g. 807 224], may be remnants of former Lower Greensand outcrops.

The basal part of the **Gault Formation** (G) occurs as partially drift-covered outliers at Whitchurch [813 203; 810 207] and Creslow [817 217; 813 216], but no evidence was found for a third outcrop [818 237] mentioned by Bristow (1963). Probably no more than 3 m of strata are present, comprising pale to dark grey, slightly silty clay with reddish brown streaks or laminae of silt. The base of the Gault is unconformable and highly transgressive. At Creslow, the Gault oversteps the Whitchurch Sand, Purbeck, and Portland formations to rest on Kimmeridge Clay within a distance of a few hundred metres. Evidently, any Lower Greensand was removed prior to deposition of the Gault. At Whitchurch, the youngest subjacent strata seen belong to the Portland Formation. The basal Gault probably belongs to the Middle Albian Dentatus Zone (Shephard-Thorn et al., 1994; Horton et al., 1995).

Structure

Palaeozoic orogenies had relatively little effect on the district. Nevertheless, the steep dips of the Lower Palaeozoic rocks in boreholes suggest deformation during pre-Late Devonian tectonic events. Regional geophysical maps suggest that these rocks are broadly disposed in a north-west trending anticline. Despite its contrary trend, the south-west trending Charlton Axis, a prominent aeromagnetic anomaly caused by magnetic igneous rocks at relatively shallow depth, may have originated at about the same time. The anomaly terminates to the north-east, where the igneous rocks are truncated by the sub-Mesozoic unconformity. A more subdued anomaly in the east may indicate similar rocks preserved on the eastern flank of the anticline.

The Upper Old Red Sandstone lies unconformably on the Lower Palaeozoic rocks. Together with the overlying Carboniferous rocks, the Upper Old Red Sandstone was subjected to Variscan folding that formed the Oxfordshire Coalfield Syncline, its axis running north–south some distance to the west of the district. An apparent offset in the syncline suggests that faults associated with the Charlton Axis were reactivated at this time.

After the Variscan episode, the district became part of the structurally stable London Platform. The gentle, south-eastward regional dip of the Mesozoic rocks (typically 0.3° to 0.5°) can probably be attributed to the

Alpine Orogeny. The Charlton Anticline in the south-west is an asymmetrical structure with a monoclinal geometry, with dips of up to 6° on the steeper north-western limb and a partly faulted crest-line (Figure 8). It is the surface manifestation of the Charlton Axis, and is probably associated with reactivated faults in these basement rocks. There is little indication that this structure persists north-eastwards from Marsh Gibbon, although a gentle, faulted anticline in the Great Oolite Group at Coombs [733 326] lies on the same trend.

Transverse faults affecting the Charlton Anticline trend towards the south-east or east-south-east, and an analogous fault has been mapped to the east of Grendon Underwood [70 20 area]. Predominantly east–west belts of faulting have been mapped on the outcrop of the Great Oolite in the north-western part of the district. The faults often occur in subparallel sets, in some cases forming narrow graben or horsts (see Section 2 on Sheet 219) in which the strata dip contrary to the regional dip direction. Throws may be 50 m or more in

places, but are generally much less. From evidence outside the district, it is likely that both fault sets were active during the early Cretaceous 'Late Cimmerian' phase.

Faults do not appear to affect more easterly parts of the Buckingham district, but minor faults are difficult to detect in the mainly mudstone succession. It is possible that faults are present there, and beneath the widespread drift cover in the north.

Superficial structures do not have a deep-seated origin, but affect only those strata near the ground surface. They are related to the development of present landforms, and probably formed during periglacial phases of the Pleistocene. These structures are most clearly seen in the south-east of the district, where the Portland and overlying strata commonly drape over the margins of the hill-tops, well below their original stratigraphical position. This condition, known as cambering, is a precursor of landslipping. The mechanism is poorly understood, but probably involves plastic deformation, outward 'flow' and subsequent wastage of mudstones of the Kimmeridge Clay, and a

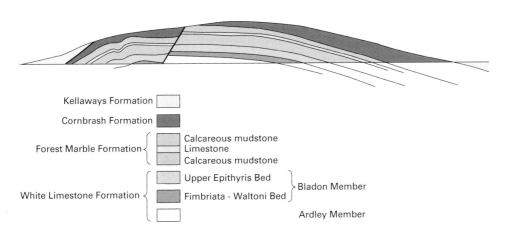

Kellaways Formation

Cornbrash Formation

Forest Marble Formation
- Calcareous mudstone
- Limestone
- Calcareous mudstone

White Limestone Formation
- Upper Epithyris Bed } Bladon Member
- Fimbriata - Waltoni Bed }
- Ardley Member

Figure 8   The Charlton Anticline as recorded by Barrow (1908) during construction of Blackthorn Hill railway cutting [6157 2122 to 6185 2092] and re-interpreted to show current stratigraphical nomenclature. Not to scale: length of section is about 360 m, thickness of strata represented is about 10 m (see Figure 4).

consequent break-up and collapse of the overlying brittle strata, which extend downslope as a 'camber' of faulted blocks. The apparent thinness of the Kimmeridge Clay at some localities is probably due to cambering. The Kimmeridge Clay on the Portland-capped Grange Hill [739 209], for example, appears to be only half its normal thickness, suggesting that this hill is an erosional remnant of a cambered spur extending from Quainton Hill to the east. In the north-west of the district, similar cambering affects the Northampton Sand Formation and Great Oolite Group along the margins of valleys incised into the Whitby Mudstone. West of Brackley, erosion has left detached cambers in several places [565 381; 561 382; 568 376].

Associated with cambering, valley bulging, which is the deformation and upward movement of mudstone strata, partly as the result of differential overburden pressure, may occur locally in valleys in the north-west. It probably also occurs on the lower slopes of the hills in the south-east, where it may explain the apparently anomalous thinning of the Ampthill Clay at some localities.

Quaternary

Deposits that may have a **preglacial** origin related to the Northern Drift Formation (the early Pleistocene deposits of the ancestral Thames river system) occur in the southern part of the district as scattered rounded pebbles of quartz and quartzite (Hey, 1986; Bowen, 1999). Alternatively, these pebbles may have been deposited immediately before the arrival of Anglian ice and thus assocatiated with the younger Hanborough Member (Sumbler, 1995, 1996). The topography beneath the Anglian deposits in the north of the district (Figure 9) suggests the presence of a valley draining north-eastwards. Gravel of locally derived limestone at the base of the drift may represent remnants of preglacial river deposits; the gravel occurs, for example, in a disused railway cutting [642 343] north of Finmere, and near Maids Moreton [715 350]. Other possible origins for the gravel include deposition

during the early stages of glaciation (compare with the Paxford Gravel of Moreton-in-Marsh; Sumbler, 2001) or scouring of local bedrock by subglacial streams.

Glacial deposits are widespread in the northern and eastern parts of the district. A thickness of more than 10 m is common-place, and over 60 m was recorded in the Deanshanger Borehole. The deposits all belong to the Wolston Formation (Bowen, 1999), which, in the district, probably relates to the second of two phases of Anglian glaciation, within oxygen-isotope Stage 10, approximately 350 000 years ago (Horton et al., 1995; Sumbler, 1995, 1996, 2001).

The most abundant glacial deposit of the district is **till**. It comprises a clayey matrix containing a variety of rock fragments (erratics). In many areas, the basal few metres of till differ from overlying deposits, being typically dark grey, sandy and silty clay with relatively few pebbles. This type of till is particularly widespread beneath the sand and gravel sheets in the north-west of the district, but 'slices' are entrained higher in the till succession in places. It is probably a ground moraine of relatively local material, incorporating preglacial soils and drift.

The greater part of the till comprises a grey, gritty, clay matrix containing erratics of chalk, flint, quartz, quartzite, Jurassic limestone, Carboniferous sandstone and limestone, and various igneous rocks. Erratics range in size from silt grade particles to large boulders. On the basis of lithology, this material may be attributed to the Oadby Till Member, which is widespread in the English Midlands. In some areas, mainly in the north-east, large 'rafts' of locally derived bedrock (mainly White Limestone or Cornbrash) are entrained.

Glaciofluvial deposits of sand and gravel are widespread. They are mainly poorly sorted flint-rich gravels and may have complex relationships with the till, with gravels variously underlying, overlying, interdigitating with or passing laterally into till. Some of the deposits rest directly on bedrock and are overlain by till. Such deposits are widespread in the Winslow area [770 277], where they are dominated

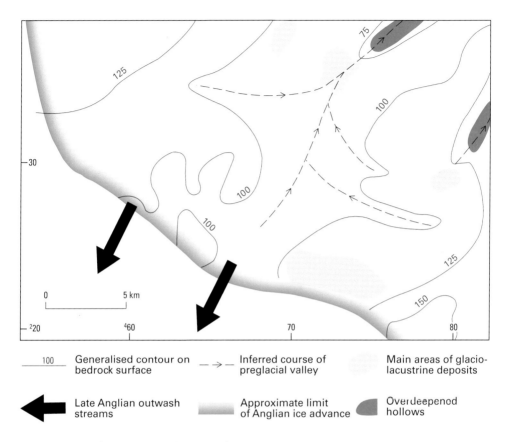

Figure 9 Rockhead topography beneath glacial deposits, assumed to approximate to the preglacial topography. The limit of ice advance is believed to have been controlled by a ridge of high ground. This bounded the Thames drainage system, into which meltwater drained during deglaciation. The two overdeepened 'tunnel valleys' have been proved by the Deanshanger and Tattenhoe boreholes (see Figures 1, 2).

by sand with relatively little gravel. They were probably laid down in front of the ice-sheet, often on an irregular preglacial surface so that lateral thickness variations may be considerable and rapid. Similar sandy deposits form a widespread sheet overlying the till to the south-west of Buckingham. They may represent an extensive outwash sandur later buried beneath advancing ice.

Many bodies of sand and gravel within the till are of limited extent; these probably include the deposits of englacial streams. Such deposits are generally ill-sorted and contain a large proportion of chalk and, in some cases, limestone. More extensive and thicker deposits of this type are present in the northern part of the district, for example in Stowe Park, where 18 m of poorly sorted gravel, containing cobbles and boulders up to 0.7 m in diameter, are exposed in a gravel pit [672 376].

Sheet-like deposits of sand and gravel overlying the till, for example at Mursley [817 286] and Steeple Claydon [703 269], were probably laid down by torrential streams as the ice-sheet melted. Probable correlative deposits form an extensive sandur in the west of the district between Finmere [636 328] and Tusmore Park [565 305]. They also extend westwards into the valley of the Ockley Brook, now a Cherwell tributary, which may have originated as an eastward-flowing feeder to the main south-westward-sloping sandur. This deposit comprises poorly sorted flint and quartz/quartzite-rich gravel.

Silty and sandy clays in which erratics are sparse dominate **undifferentiated glacial deposits** on Sheet 219. Such material, usually closely associated with till or sand and gravel, was probably deposited in lakes, either on, in front of, or beneath the ice. The most extensive mapped deposits of this type cap the hill [723 227] between Botolph Claydon and Shipton Lee, and similar deposits occur on Knowl Hill [702 241] to the north-west. Their situation (Figure 9) suggests deposition from meltwater ponded against higher ground (now mostly eroded away) to the south. Other deposits of this type occur near Tingewick [651 333], Horwood House [793 293] and most extensively in the Akeley–Foscote–Thornton–Leckhampstead area to the north and north-east of Buckingham. There, the deposits are believed to be associated with a deep subglacial channel or tunnel-valley, more or less coincident with the present Great Ouse valley (Figure 9), and may therefore be widespread beneath a cover of younger drift in this area. This was confirmed by the Deanshanger Borehole just outside the district, which proved over 60 m of mainly glaciolacustrine silts and clays, their base only 6 m above sea level. This is presumed to represent localised over-deepening of the preglacial valley by pressurised meltwater beneath the ice-sheet. Similar though less dramatic deepening in a tributary valley on the eastern margin of the district (Figure 9) is indicated by a 39 m succession of till and glaciolacustrine deposits in the Tattenhoe Borehole.

Fluvial deposits include alluvium and river terrace deposits. In the northern part of the district, **River Terrace Deposits** are developed most strongly along the River Great Ouse downstream from Buckingham. The deposits are typically flint and quartz/quartzite rich gravels, with variable amounts of limestone. They are composed largely of reworked glacial deposits and may therefore be similar in composition to the latter. However, they are generally better sorted, with a smaller proportion of fines and few non-durable clasts such as chalk or mudstone. Two phases of terrace aggradation can be distinguished in places, but numbering is of purely local significance. Their relationship to the better-studied terrace succession farther downstream along the Great Ouse or in the Thames catchment is uncertain. The Second Terrace Deposits form a bench up to 5 m above the modern floodplain, while the younger First Terrace Deposits form a low, sloping bench up to about 2 m above the floodplain, with the gravels continuing beneath the alluvium.

Substantial terrace deposits also occur along the Padbury and Claydon brooks, the principal south bank tributaries of the Great Ouse, with minor deposits present in streams elsewhere. These deposits are typically less well sorted than those along the main river, comprising loamy gravels with pebbles up to 10 cm in diameter. Mostly they form gently sloping or mounded deposits extending up to perhaps 7 m above the modern floodplain. Their distribution suggests that they represent the dissected and partly reworked remnants of valley fills.

Patches of river terrace deposits in the southern part of the district were laid down along Thames tributaries. In the south-east corner of the district [817 194], a small patch of stony loam forms a terrace adjoining the Hardwick Brook. More extensive terrace deposits occur along the Langford Brook to the south-west of Bicester.

Deposits classified as Second Terrace cap low hills near Little Chesterton [554 200; 563 208]. They comprise limestone-rich gravels, leached and oxidised near surface to produce an almost pebble-free loam soil. The deposits probably equate with part of the composite Summertown–Radley Member of the Thames succession (see Sumbler, 1995), which is no younger than Early Devensian. Deposits assigned to the First Terrace form poorly defined terraces south-east of Chesterton [573 205], rising up to about 1.5 m above the adjoining floodplain. The deposits comprise well-sorted, fine grade, limestone-rich gravel, covered by a humic, slightly sandy loam. The gravels extend beneath the alluvium; thicknesses of up to 3 m are probable. They probably equate with the composite Northmoor Terrace of the Thames succession (see Sumbler, 1995) of Devensian age.

Almost all the larger streams and rivers of the district have a flat alluvial floodplain. The **alluvium** seldom exceeds 2 to 2.5 m in thickness, and is typified by clay, sandy clay and silt. It commonly has a thin gravelly deposit at the base, representing the coarser lag material from the bottom of channels, but the thicker sand and gravel deposits present beneath the floodplains of the Great Ouse and Langford Brook are probably contemporaneous with the youngest river terrace deposits.

Extensive alluvium occurs in the valley of the River Ray on the southern margin of the district. The deposits consist largely of re-worked Jurassic mudstone from the local bedrock, with a component of humic, peaty material near the surface and gravelly pods and lenses containing flint and other pebbles at depth. The alluvium along the Langford Brook and its tributaries is simi-larly composed of dark humic and locally peaty clay and silt, commonly highly cal-careous. It is generally underlain by a more or less sandy gravel of local limestone pebbles. The alluvial clays and silts of the Great Ouse and its tributaries are commonly very calcareous, reflecting derivation from Middle Jurassic limestone terrain. They also contain a substantially greater proportion of material derived from glacial deposits, which are widespread within the catchment.

Head is a solifluxion deposit that proba-bly dates mainly from periglacial phases of the Quaternary, when the presence of per-mafrost led to waterlogging and resultant mobilisation of the surface layers during summer thaws. Head deposits are typically poorly sorted, more or less stony clays (diamictons) made up of reworked material from the hill slopes. They may be difficult to distinguish from the parent material, partic-ularly where this is till or mudstone, and consequently only lithologically or morpho-logically distinct deposits are indicated on the map. The latter typically form fans or gently sloping 'terraces' at the foot of hill slopes, or infill valley bottoms.

Extensive **landslips** affect the hills in the south-east of the district, where slopes of Kimmeridge Clay are capped by the Portland, Purbeck and Whitchurch Sand formations. Groundwater seepage has probably weak-ened the mudstone and lubricated potential slip-surfaces. The upper slopes of Quainton Hill [75 21] show a stepped topography resulting from multiple, small-scale rotational slips (Plate 1). Lobate fans formed by mud-flows occur on the lower slopes. Mapped areas of landslip are relatively small on the Oving–Whitchurch hills, to the east, but it is possible that some slips have been obscured by agricultural improvements. A few small areas of landslip have been mapped in other parts of the district, generally where seepage of water from permeable drift deposits has weakened slopes of mudstone or till.

Small areas of **tufa** and **peat** have been mapped, usually in close association with each other.

Artificially modified ground, has been substantially altered by man's activities, either through excavation (mainly quarry-ing), by tipping, or by a combination of the two. For reasons of scale, only the largest such areas are shown on the map. In many cases, the boundaries shown may be approx-imate, due to the difficulty of delineating the limits of former disturbance once restoration is complete. For similar reasons, the depiction of landscaped ground in urban areas is not comprehensive. Artificial road and railway cuttings are indicated on the topographical basemap.

3   Applied geology

Engineering geology

Overconsolidated clays, mud-rocks that have been compacted as a result of burial, are extensive. They include the Whitby Mudstone, virtually the whole of the Ancholme Group, and thin units in the Great Oolite, Portland and Purbeck groups. These form clay-rich, cohesive soils, which have a relatively low strength near surface because of weathering and stress release, but increased strength with depth, concomitant with a reduction in moisture content. They are commonly of moderate to high plasticity. Very high plasticities may be encountered in the Ampthill Clay and West Walton formations, in clay beds within the Bladon Member, the Forest Marble, Purbeck, Whitchurch Sand and Gault formations, and in glaciolacustrine deposits. Glacial till also generally has a clay-rich matrix, but varies considerably in composition and consequently in geotechnical properties.

Normally consolidated deposits tend to have low or very low strength. Alluvium is the principal example in the district, but head and some landslip deposits may have similar properties and may also contain relict shear surfaces. High moisture contents are typical, and plasticity may range to extremely high values.

Most of the potential site-engineering problems are associated with the clay formations, and appropriate precautions must be taken, for example when excavating in plastic or sheared clays. Extensive areas of landslip particularly affect the Kimmeridge Clay, and care should be taken to avoid reactivating slips by overloading, undercutting or inappropriate drainage works. Any clay slope steeper than about 7° to 10° may be potentially unstable (Horton et al., 1995). Shrinking and swelling with changing moisture content is likely to be a problem in the more highly plastic clays near surface,

and the possibility of sulphate attack on buried concrete should be considered in connection with pyrite-bearing clays such as those of the Whitby Mudstone Formation and Ancholme Group.

Of the non-cohesive soils, sands occur in the Horsehay Sand and Whitchurch Sand formations and the Kellaways Sand and Portland Sand members, and gravels are mainly represented by glaciofluvial and river terrace deposits. Sands and gravels are dense at depth and may be cemented locally, but most become loose near the surface. Excavations below the water table in sandy materials may encounter running sand conditions, and appropriate measures must be taken to control water inflow.

The rocks are mainly limestones such as those of the Great Oolite, Portland and Purbeck groups, with a few sandstones, for example in the Northampton Sand Formation and lenses within the sand formations listed above. At depth they are generally weak to moderately strong, but most are broken-up in the near-surface zone as a result of weathering, and bedding and joint planes are dilated. In certain situations, they may be affected by cambering, with the possible presence of major subsurface fissures and voids, some of which may be partly infilled with soil.

Most formations contain thin beds of contrasting material, which may have an important influence on the bulk properties. For this reason, any map-based desk study of the geology prior to major excavation or construction work should be followed by a detailed on-site investigation.

Artificial fill has a varied and unpredictable composition, and may involve additional hazards such as methane release. The Northampton Sand is associated with radon release elsewhere (Appleton and Ball, 1995), so it is possible that raised levels of radon may occur in soils and foundations in the north-west part of the district.

Hydrogeology and water resources

Groundwater is abstracted from a number of sources, mainly for agricultural and industrial use. Several boreholes were used for public water supply, but have now been replaced by larger sources outside the district. Many of the higher yielding boreholes are open to more than one aquifer. The principal aquifer is the Great Oolite Group, but minor aquifers occur throughout the Jurassic, in the Lower Cretaceous, and in the superficial deposits.

The Northampton Sand Formation (Inferior Oolite Group) and Horsehay Sand Formation (Great Oolite Group) are in hydraulic continuity and together form a good aquifer, yielding water of good quality. At higher stratigraphical levels in the Great Oolite Group, the White Limestone Formation gives high yields, whereas the Cornbrash Formation contains limited supplies giving low yields; water levels in the latter were originally artesian near Launton Station [6219 2350] and still are at Blackstone Farm, Blackthorn [627 201].

The Oxford Clay, West Walton, Ampthill Clay and Kimmeridge Clay formations act together as aquicludes, confining water in the underlying Great Oolite. Small supplies of shallow groundwater have been obtained in the Marsh Gibbon area, from wells sited on the Peterborough Member that may penetrate the Kellaways Formation. The Portland and Whitchurch Sand formations provide small quantities of groundwater in the southeast of the district, mainly from springs.

Small-scale domestic supplies have been obtained from sand and gravel lenses within the till, but the main hydrogeological significance of the till is that it reduces recharge into the underlying formations. Glaciofluvial sand and gravel deposits commonly occur as small isolated outcrops with limited recharge. Supplies are likely to be affected by seasonal variations in water level, but were once utilised for small-scale domestic purposes. Water quality is often similar to that of the underlying formations. Sands and gravels of the river terrace deposits are generally in hydraulic continuity with a river, so yields are more likely to be sustained than

from other sources in the superficial deposits. However, the groundwater is shallow and liable to surface pollution.

Shallow groundwater is highly vulnerable to contamination from both diffuse (e.g. nitrates and pesticides) and point source (e.g. storage tanks) pollutants, due to the thinness of the unsaturated zone. Successful aquifer remediation is difficult, prolonged and expensive, and therefore the prevention of pollution is important. Water in the deeper Great Oolite aquifer is often confined and well protected from pollution.

Mineral and energy resources

Stone from the Great Oolite Group is used in the many stone buildings in the west of the district. Most of the freestone used for quoins and mullions is Chipping Norton Limestone from the district to the west, but the local Cornbrash has been used extensively for rough walling, and stone from the Taynton Limestone and White Limestone formations has also been used. The latter is not especially durable, being dominated by micritic, frost-susceptible lithologies. Stone from the Portland and Purbeck formations is an attractive feature of many buildings in the southeast of the district, but suffers from the same durability problem. It is unlikely that there are any viable building stone resources in the district, but the White Limestone is worked on a large scale for aggregate and fill near Ardley [543 259], and was also worked until relatively recently at Croughton [535 336] and Stratton Audley [601 252]. Formerly, the Cornbrash limestone was widely quarried for roadstone and aggregate.

The sandy uppermost beds of the Kimmeridge Clay were once worked for brick, pipe and tile manufacture at Whitchurch [807 204]. On a larger scale, the clays of the Peterborough Member of the Oxford Clay were worked for brick manufacture at Calvert [685 246] for nearly a century, prior to closure in 1991. The old pits, which cover more than 200 hectares, are used variously as a sailing lake, nature reserve and landfill. The Peterborough

Member is highly suitable for brick manufacture, and large reserves remain in the district.

Substantial deposits of sand and gravel occur in the northern part of the district (Sumbler, 1990, 1991). The glaciofluvial deposits are likely to contain a high proportion of fines and non-durable lithologies such as chalk, which may affect their commercial viability. Nevertheless, such deposits are currently worked near Finmere [628 324], and prospects near Mursley and Chackmore have been considered in the past. More promising are the river terrace gravels of the Great Ouse. These have been worked near Foscote [725 352] and it is likely that reserves remain downstream from there, partly beneath the river's floodplain.

Natural gas was encountered during drilling of a water well at Calvert in 1905.

The possibility of a hydrocarbon reservoir was investigated in the 1960s and 1970s, but no economic quantities of gas were found. Drilling of many deep boreholes in the western part of the district and beyond, in connection with an abortive scheme to store imported natural gas in the Triassic sandstone 'reservoir' (see Horton et al., 1987), revealed the limits of the Carboniferous rocks at depth.

The southernmost of the Calvert Brickpits [697 230] is used as a major landfill site. Domestic and other waste is transported to the site by road and rail and is buried in carefully constructed and monitored pounds, with facilities for leachate treatment and methane utilisation. On a smaller scale, landfill is taking place at Finmere gravel pit [628 324], Ardley Quarry [543 259] and Stratton Audley Quarry [601 252].

Information sources

Further sources of geological information relevant to the district are listed below, although they may be superseded by this account. Other sources include unpublished internal reports, borehole records, hydrogeological data, photographs and material collections such as fossils, rock samples and thin sections. Indexes to some of the collections can be searched on the BGS Internet Geoscience Data Index (accessible through the BGS web site) or by personal visit to a BGS office. BGS can also carry out searches of these data on your behalf. Please contact the BGS Central Enquiry Desk for further details.

Maps

A range of small-scale geological, hydrogeological, geophysical and geochemical maps are available. Please consult current BGS catalogues for details. Only the larger scale geological maps are listed below.

1:25 000
Milton Keynes, Solid and Drift, 1971

1:50 000 and 1:63 360
Sheet 201 Banbury, Solid and Drift, 1982
Sheet 202 Towcester, Solid and Drift, 1969
Sheet 203 Bedford. Contact BGS Central
    Enquiry Desk for further details.
Sheet 218 Chipping Norton, Solid and
    Drift, 1968
Sheet 219 Buckingham, Solid and
    Drift, 2002
Sheet 220 Leighton Buzzard, Solid and
    Drift, 1992
Sheet 236 Witney, Solid and Drift, 1982
Sheet 237 Thame, Solid and Drift, 1994
Sheet 238 Aylesbury, Solid and Drift, 1923

1:10 000
The Buckingham district has a complete coverage of 1:10 000 scale geological sheets. The original hand-drawn maps may be consulted at the BGS Library, Keyworth, or uncoloured dyeline copies may be purchased.

Books

British Regional Geology: London and the Thames Valley. Fourth edition, 1996

Memoirs
Geology of the country around Banbury and Edge Hill (Sheet 201), 1965
Geology of the country around Chipping Norton (Sheet 218), 1987
Geology of the country around Leighton Buzzard (Sheet 220), 1994
Geology of the country around Witney (Sheet 236), 1946
Geology of the country around Thame (Sheet 237), 1995
Geology of the country around Aylesbury and Hemel Hempstead (Sheet 238), 1922

Geological and Mineral Assessment Reports
The geology of the New Town of Milton Keynes: explanation of 1:25 000 special geological sheet SP83 with parts of SP73, 74, 84, 93, and 94. *Report of the Institute of Geological Sciences*, No. 74/16 (1974).

A preliminary assessment of the sand and gravel deposits of part of the Ouse Valley in Bedfordshire, Buckinghamshire and Northamptonshire (1:25 000 sheets SP 84, 85, 95 and TL 05): *Report of the Institute of Geological Sciences*, WF/MN/82/7 (1982).

A preliminary study of the sand and gravel and other potential mineral deposits of parts of north Buckinghamshire, south Northamptonshire and east Oxfordshire (1:25 000 sheets SP 62, 63, 64, 72, 73, 74, 82, and parts of SP 52, 53, 54, 55, 65 and 75). Report of the Institute of Geological Sciences, WF/MN/84/6 (1983).

Sand and gravel deposits near Buckingham. *British Geological Survey Technical Report*, WA/90/71 (1990).

A preliminary study of potential resources of sand and gravel in Buckinghamshire north of the Chilterns. *British Geological Survey Technical Report*, WA/90/50 (1990).

Sand and gravel deposits west of Buckingham. *British Geological Survey Technical Report*, WA/91/21 (1991).

Geology of the Bicester area (SP 52 SE). *British Geological Survey Technical Report*, WA/01/10 (2001).

Geology of the Mursley area (SP 82 SW). *British Geological Survey Technical Report*, WA/01/12 (2001).

Geology of the Brackley and Charlton area (SP 53 NE and part of SP 53 NW). *British Geological Survey Internal Report*, IR/01/081 (2001).

Documentary collections

Boreholes

Borehole and well records for the district are catalogued in the National Geological Records Centre (BGS Keyworth), and at the National Well Records Archive (BGS Wallingford). Please contact the BGS Central Enquiry Desk for further details.

BGS Lexicon of named rock units

Definitions of the named rock units shown on BGS maps, including those shown on 1:50 000 Sheet 219 Buckingham, are held in the Lexicon database, available on the BGS web site. Further information on the database can be obtained from the Lexicon Manager at BGS Keyworth

BGS photographs

Copies of the photographs listed below are deposited for reference in the BGS Library, Keyworth. Those prefixed with BAAS derive from the British Association collection lodged with BGS.

A3220–A3225 Oxford Clay Formation, Calvert Brickyard, 1925
A11203–A11204 Glacial drift deposits, Whaddon, 1968
BAAS05333–BAAS05335 Great Oolite, Blackthorn Hill railway cutting, 1908

References

Most of the references listed below are held in the libraries of the British Geological Survey at Murchison House, Edinburgh and at Keyworth, Nottingham. Copies of the references can be purchased subject to the current copyright legislation.

ANDERSON, F W. 1985. Ostracod faunas in the Purbeck and Wealden of England. *Journal of Micropalaeontology*, Vol. 4, 1–67.

APPLETON, J D, and BALL, T K. 1995. Radon and background radioactivity from natural sources: characteristics, extent and relevance to planning and development in Great Britain. *British Geological Survey Technical Report*, No. WP/95/2.

ARKELL, W J, RICHARDSON, L, and PRINGLE, J. 1933. The Lower Oolites exposed in the Ardley-Fritwell railway cuttings, between Bicester and Banbury, Oxford. *Proceedings of the Geologists' Association*, Vol. 44, 340–54.

BARKER, D. 1966. Ostracods from the Portland and Purbeck Beds of the Aylesbury District. *Bulletin of the British Museum (Natural History), Geology*, Vol. 11, 458–487.

BARROW, G. 1908. The new Great Central Railway from Ashendon to Aynho near Banbury. *Summary of Progress of the Geological Survey of Great Britain* (for 1907), 141–154.

BOWEN, D Q. (editor). 1999. A revised correlation of Quaternary deposits in the British Isles. *Geological Society of London Special Report*, No. 23.

BRADSHAW, M J. 1978. A facies analysis of the Bathonian of eastern England. Unpublished DPhil. thesis, University of Oxford.

BRISTOW, C R. 1963. The stratigraphy and structure of the Upper Jurassic and Lower Cretaceous rocks in the area between Aylesbury (Bucks) and Leighton Buzzard (Beds.). Unpublished PhD Thesis, University of London.

BRISTOW, C R. 1968. Portland and Purbeck Beds. 300–311 in *The geology of the East Midlands*. SYLVESTER-BRADLEY, P C and FORD, T D (editors). (Leicester: Leicester University Press.)

BRISTOW, C R, and KIRKALDY, J F. 1962. Field meeting in the Leighton Buzzard-Aylesbury area. *Proceedings of the Geologists' Association*, Vol. 73, 455–459.

BULMAN, O M B, and RUSHTON, A W. 1973. Tremadoc faunas from boreholes in Central England. *Bulletin of the Geological Survey of Great Britain*, No. 43, 1–40.

BUTLER, D E. 1981. Marine faunas from concealed Devonian rocks of southern England and their reflection of the Frasnian transgression. *Geological Magazine*, Vol. 118, 679–697.

CALLOMON, J H. 1968. The Kellaways Beds and the Oxford Clay. 264–290 in *The geology of the East Midlands*. SYLVESTER-BRADLEY, P C and FORD, T D (editors). (Leicester: Leicester University Press.)

CASEY, R, and BRISTOW, C R. 1964. Notes on some ferruginous strata in Buckinghamshire and Wiltshire. *Geological Magazine*, Vol. 101, 1160–128.

COX, B M, HOPSON, P M, and SUMBLER, M G. 1991. A new record of the Upper Cornbrash near Buckingham. *Proceedings of the Geologists' Association*, Vol. 102, 63-65.

COX, B M, HUDSON, J D, and MARTILL, D M. 1992. Lithostratigraphic nomenclature of the Oxford Clay (Jurassic). *Proceedings of the Geologists' Association*, Vol. 103, 343–345.

COX, B M, and SUMBLER, M G. 2002. British Middle Jurassic stratigraphy. *Geological Conservation Review Series*, No. 26. (Peterborough: Joint Nature Conservation Committee.)

DAVIES, A M. 1897. Excursion to Whitchurch, Oving and Quainton. *Proceedings of the Geologists' Association*, Vol. 15, 207–209.

DAVIES, A M, and PRINGLE, J. 1913. On two deep borings at Calvert Station (North Buckinghamshire). *Quarterly Journal of the Geological Society of London*, Vol. 69, 308–340.

DONOVAN, D T, HORTON, A, and IVIMEY-COOK, H C. 1979. The transgression of the Lower Lias over the northern flank of the London Platform. *Journal of the Geological Society of London*, Vol. 136, 165–173.

DOUGLAS, J A, and ARKELL, W J. 1932. The stratigraphical distribution of the Cornbrash: II. The north-eastern area. *Quarterly Journal of the Geological Society of London*, Vol. 88, 112–170.

FENTON, J P G, RIDING, J B, and WYATT, R J. 1994. Palynostratigraphy of the Middle Jurassic 'White Sands' of central England. *Proceedings of the Geologists' Association*, Vol. 105, 225–230.

FENTON, J P G, RIDING, J B, and WYATT, R J. 1995. 'Palynostratigraphy of the Middle Jurassic "White Sands" of central England' by Fenton, Riding and Wyatt (1994): reply. *Proceedings of the Geologists' Association*, Vol. 106, 306–308.

FITTON, W. 1836. Observations on some strata between the Chalk and the Oxford Oolite in the southeast of England. *Transactions of the Geological Society of London*, Vol. 36, 189–236.

FOSTER, D, HOLLIDAY, D W, JONES, C M, OWENS, B, and WELSH, A. 1989. The concealed Upper Palaeozoic rocks of Berkshire and South Oxfordshire. *Proceedings of the Geologists' Association*, Vol. 100, 395–407.

GREEN, A H. 1864. The geology of the country around Banbury, Woodstock, Bicester and Buckingham. *Memoir of the Geological Survey of Great Britain*. [Old Series] Sheet 45.

HEY, R W. 1986. A re-examination of the Northern Drift of Oxfordshire. *Proceedings of the Geologists' Association*, Vol. 97, 291–302.

HORTON, A, POOLE, E G, WILLIAMS, B J, ILLING, V C, and HOBSON, G D. 1987. Geology of the country around Chipping Norton. *Memoir of the British Geological Survey*. Sheet 218 (England and Wales).

HORTON, A, SHEPHARD-THORN, E R, and THURRELL, R G. 1974. The geology of the new town of Milton Keynes. Explanation of 1:25 000 Special Geological Sheet SP 83 with parts of SP 73, 74, 84, 93 and 94. *Report of the Institute of Geological Sciences*, No. 74/16.

HORTON, A, SUMBLER, M G, COX, B M, and AMBROSE, K. 1995. Geology of the country around Thame. *Memoir of the British Geological Survey*, Sheet 237 (England and Wales).

MITCHELL, J. 1834. The strata of Quainton and Brill in Buckinghamshire. *Proceedings of the Geological Society of London*, Vol. 2, 6–7.

MORTER, A A. 1984. Purbeck-Wealden Beds Mollusca and their relationship to ostracod biostratigraphy, stratigraphical correlation and palaeoecology in the Weald and adjacent areas. *Proceedings of the Geologists' Association*, Vol. 95, 217–234.

ODLING, M. 1913. The Bathonian rocks of the Oxford district. *Quarterly Journal of the Geological Society of London*, Vol. 69, 484–513.

PALMER, T J. 1979. The Hampen Marly and White Limestone formations: Florida-type carbonate lagoons in the Jurassic of central England. *Palaeontology*, Vol. 22, 189–228.

PHARAOH, T C, MERRIMAN, WEBB, P C, and BECKINSALE, R D. 1987. The concealed Caledonides of eastern England: preliminary results of a multidisciplinary study. *Proceedings of the Yorkshire Geological Society*, Vol. 46, 355–369.

PHARAOH, T C, MERRIMAN, R J, EVANS, J A, BREWER, T S, WEBB, P C, and SMITH, N J P. 1991. Early Palaeozoic arc-related volcanism in the concealed Caledonides of southern Britain. *Annales de la Société Géologique de Belgique*, Vol. 114, 63–91.

POOLE, E G. 1969. The stratigraphy of the Geological Survey Apley Barn Borehole, Witney, Oxfordshire. *Bulletin of the Geological Survey of Great Britain*, No. 29, 1–103.

POOLE, E G. 1977. The stratigraphy of the Steeple Aston Borehole, Oxfordshire. *Bulletin of the Geological Survey of Great Britain*, No. 57.

POOLE, E G. 1978. Stratigraphy of the Withycombe Farm Borehole, near Banbury, Oxfordshire. *Bulletin of the Geological Survey of Great Britain*, No. 68.

RUSHTON, A W A, and MOLYNEUX, S G. 1990. The Withycombe Formation (Oxfordshire subcrop) is of early Cambrian age. *Geological Magazine*, Vol. 127, 363.

SELLWOOD, B W, and McKERROW, W S. 1974. Depositional environments in the lower part of the Great Oolite Group of Oxfordshire and north Gloucestershire. *Proceedings of the Geologists' Association*, Vol. 85, 189–210.

SHEPHARD-THORN, E R, MOORLOCK, B S P, COX, B M, ALLSOP, J M, and WOOD, C J. 1994. Geology of the country around Leighton Buzzard. *Memoir of the British Geological Survey*, Sheet 220 (England and Wales).

SUMBLER, M G. 1984. The stratigraphy of the Bathonian White Limestone and Forest Marble formations of Oxfordshire. *Proceedings of the Geologists' Association*, Vol. 95, 51–64.

SUMBLER, M G. 1989. Geological notes and local details for 1:10 000 Sheet SP 71 NE (Fleet Marston). *British Geological Survey Technical Report*, WA/89/32.

SUMBLER, M G. 1990. Sand and gravel deposits near Buckingham. *British Geological Survey Technical Report*, WA/90/71.

SUMBLER, M G. 1991.   Sand and gravel deposits west of Buckingham.   *British Geological Survey Technical Report*, WA/91/21.

SUMBLER, M G. 1995.   The terraces of the rivers Thame and Thames and their bearing on the chronology of glaciation in central and eastern England.   *Proceedings of the Geologists' Association*, Vol. 106, 93–106.

SUMBLER, M G. 1996.   *British Regional Geology: London and the Thames Valley.* Fourth Edition. (London: HMSO.)

SUMBLER, M G. 2001.   The Moreton Drift; a further clue to glacial chronology in central England.   *Proceedings of the Geologists' Association*, Vol. 112, 13–27.

SUMBLER, M G, and BARRON, A J M. 1996.   The type section of the Hampen Formation (Middle Jurassic, Great Oolite Group) at Hampen Cutting, Gloucestershire.   *Proceedings of the Cotteswold Naturalists' Field Club*, Vol. 41, 118–28.

WHITAKER, W. 1921.   The water supply of Buckinghamshire and of Herefordshire from underground sources.   *Memoir of the Geological Survey of England and Wales.*

WIMBLEDON, W A. 1974.   The stratigraphy and ammonite faunas of the Portland Stone of England and northern France. Unpublished PhD. Thesis, University of Wales.

WIMBLEDON, W A. 1980.   Portlandian correlation chart.   85–93 *in* A correlation of the Jurassic rocks of the British Isles. Part Two: Middle and Upper Jurassic. COPE, J C W (editor).   *Geological Society of London Special Report*, No. 15.

WOODWARD, H B. 1895.   The Jurassic rocks of Britain. Vol. 5. The Middle and Upper Oolitic rocks of England (Yorkshire excepted). *Memoir of the Geological Survey of the United Kingdom.*